T0132883

FLORA OF TROPICAL EAST AFRICA

PLUMBAGINACEAE

C. M. WILMOT-DEAR

Perennial, rarely annual, herbs or small shrubs. Leaves alternate or in basal rosettes, exstipulate. Flowers 1–several in usually 3-bracteate spikelets; spikelets grouped into spikes or compact heads. Sepals united, tubular or funnel-shaped, 5-nerved, often 5-ribbed or 5-angled; limb sometimes membranous or scarious. Corolla actinomorphic, tubular or funnel-shaped with 5 lobes, or petals connate only at base. Stamens 5, antipetalous, inserted at base of corolla; anthers dithecous, dehiscent longitudinally. Ovary superior, sessile or stalked, 1-locular; ovule 1, anatropous, pendulous from a basal funicle; styles 5, or style 1 with 5 stigma-lobes. Fruit a dry 1-seeded capsule, often enclosed in the persistent calyx, indehiscent or dehiscing irregularly or by splitting in a complete ring near base, or operculate. Endosperm abundant, scanty or absent.

A cosmopolitan family of 17 genera, especially in arid, saline inland and maritime habitats.

1. Calyx with spreading scarious limb; styles 5, distinct   1. **Limonium**
   Calyx with erect lobes; style 1 with 5 stigma-lobes   . .   2
2. Flowers in dense terminal or lateral heads; calyx
   eglandular . . . .   2. **Ceratostigma**
   Flowers in lax to dense ± elongated spikes; calyx
   glandular . . . .   3. **Plumbago**

1. LIMONIUM

Mill., Gard. Dict., abridg. ed. 4, 2 [unpaged] (1754), *nom. conserv.*

Statice sensu auctt., *non* L. (1753)

Perennial, rarely annual, herbs or small shrubs. Inflorescence paniculate, often corymbose; spikes secund, sometimes grouped into compact terminal heads; spikelets 1–several-flowered, 2–several-bracteate, with a small fleshy bracteole at the base of each bract. Calyx funnel-shaped, 5-nerved; limb scarious, 5-lobed. Corolla with short tube or petals connate only at base. Stamens inserted at base of corolla. Styles 5, glabrous, free or connate at base; stigma filiform. Fruit a capsule, membranous, 1-seeded, enclosed in persistent calyx, indehiscent or irregularly dehiscent or operculate. Seed dark brown or black, long, narrow, slightly flattened, tapering slightly towards apex.

An almost worldwide genus of about 300 species, especially well developed in Central Asia and the Mediterranean region.

Several species are, or have been, cultivated as ornamental plants, among which the three following have been identified.

L. perezii (Stapf) L. H. Bailey (*Statice perezii* Stapf), native to the Canary Islands, has been grown at Nairobi, e.g. Oct. 1961, *Graham Bell* 10! & Nov. 1961, *Graham Bell* in

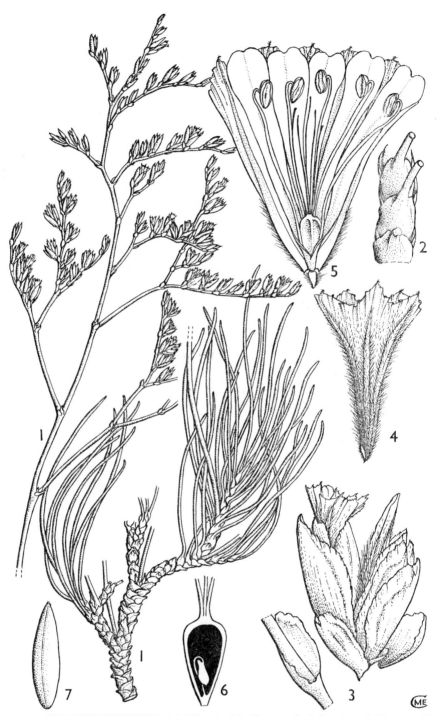

FIG. 1. *LIMONIUM DISTICHUM*—**1,** habit, × ⅔; **2,** leaf-bases, × 2; **3,** flowering spikelet, × 6, with detail of bracteole and peduncle from within; **4,** calyx, × 6; **5,** flower, opened out, × 9; **6,** longitudinal section of ovary, × 20; **7,** seed, × 14. All from *IECAMA* 1–86.

E.A.H. 318/61B !. Perennial herb, with long-petioled rhomboid leaves and corymbose inflorescence of 2-flowered spikelets; calyx-tube 5–6 mm. long, lilac limb 4 mm. long and yellowish corolla.

L. sinuatum Mill., native to the Mediterranean region, has been grown near Nairobi, e. g. Langata, 16 Sept. 1972, *Nzoika* N. 3 !. Perennial herb, bearing narrow crisp wings along stem with leaf-like extensions at nodes, pinnatifid leaves with alternate rounded lobes, compact terminal inflorescences of 3-flowered spikelets, calyx 6–7 mm. long with wide purple limb, and yellow or pink corolla.

L. vulgare Mill., native to Europe and North America, or a hybrid of this species, has been cultivated near Nairobi—Nov. 1961, *Graham Bell* in *E.A.H.* 318/61A !; exact identification uncertain. Perennial herb, with oblong or obovate leaves, corymbose inflorescence of 2-flowered spikelets, corolla-tube and calyx both 4 mm. long, and reddish corolla-lobes.

1. Leaves pinnatifid, with 6–10 rounded lobes (cultivated) . *L. sinuatum*
   Leaves entire 2
2. The leaves cylindrical, fleshy, 1–2 mm. in diameter . *L. distichum*
   The leaves broader, flattened, oblong, rhombic or obovate,
   never cylindrical (cultivated) 3
3. Leaves rhombic, mucronate, with sparse fine hairs above
   and beneath; primary venation pinnate . . . *L. perezii*
   Leaves oblong or obovate, rounded, glabrous; primary
   venation not pinnate . . . . *L. vulgare*

L. distichum *Wilmot-Dear* in K.B. 31 (1976). Type: Somalia, mouth of Nogal, Eil, *Hemming* 1676 (K, holo. !, EA, iso.)

Perennial herb or shrub, 15–50 cm. high. Stems stout, woody, often completely covered by old hard dry overlapping amplexicaul bases of shed leaves. Leaves fleshy, cylindrical, (30–)40–70 mm. long, 1–2 mm. in diameter, rugulose, glabrous, mucronate with mucro 0·5 mm. long, sessile, distichous, alternate, bases amplexicaul and overlapping, completely covering stem. Inflorescence paniculate; spikelets alternate, fairly laxly arranged, 3(rarely–4)-flowered, 5-bracteate; bracts broadly obovate or elliptical, 5–8 mm. long, 3–5 mm. broad, with broad or narrow hyaline margins, obtuse, glabrous. Calyx 7–8 mm. long, 1 mm. wide, both tube and limb often densely pubescent; tube often completely enclosed by bracts; limb white, with somewhat irregular margin. Corolla blue; petals connate at base, 7 mm. long, 1 mm. wide. Styles 5. Fig. 1.

KENYA. Northern Frontier Province: 32 km. W. of Mandera, on border road to Moyale, 27 Sept. 1953, *Bally* 9331 !
DISTR. **K**1; Ethiopia, Somali Republic
HAB. *Euphorbia* bushland; ± 250 m.

2. CERATOSTIGMA

Bunge, Enum. Pl. Chin. Bor.: 55 (1833)

Valoradia Hochst. in Flora 25: 239 (1842)

Perennial herbs or small shrubs. Leaves alternate, simple. Flowers in 3-bracteate 1–2-flowered spikelets grouped into compact terminal and axillary heads; central (outer) bract largest. Calyx tubular, 5-ribbed, somewhat scarious. Corolla tubular, with patent limb. Stamens free. Style 1, with 5 stigma-lobes. Fruit a capsule, membranous, 1-seeded, enclosed in the persistent calyx, dehiscent by splitting into 5 valves at base. Seed dark brown, narrowly ovoid, tapering markedly towards tip, slightly flattened, surface covered with minute stellate protuberances.

A small genus of about 8 species, mostly in China and the Himalayas, one in tropical Africa.

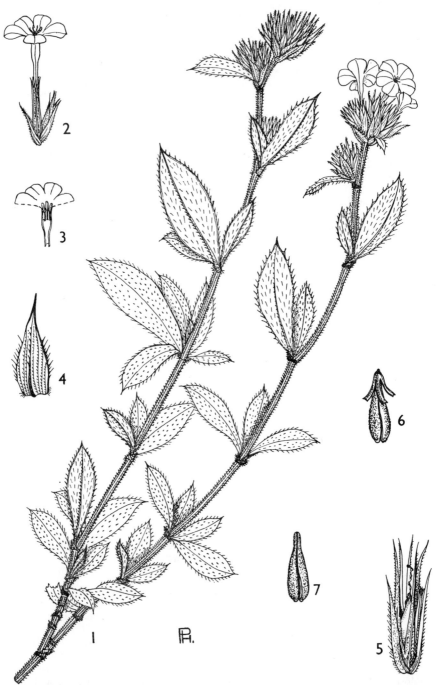

FIG. 2. *CERATOSTIGMA ABYSSINICUM*—**1,** habit, × ⅔; **2,** flowering spikelet, × 1; **3,** corolla opened out to show stamens, × 1; **4,** outer bract, × 2; **5,** fruiting spikelet, × 2; **6,** seed attached to remains of fruit, × 3; **7,** seed, × 3. 1, 5, from *Bally & Smith* 14882; 2–4, from *Tweedie* 4301; 6, 7, from *Gillett* 13789.

C. abyssinicum (*Hochst.*) *Schweinf. & Aschers.* in Schweinf., Beitr. Fl. Aeth.: 288 (1867); Prain in J.B. 44: 8 (1906); E.P.A.: 660 (1960); U.K.W.F.: 508 (1974). Type: Ethiopia, Mt. Scholoda, *Schimper* 253 (K, iso.!)

Shrub or herb, often low and tangled. Stems rigid, bent slightly at nodes in a zigzag manner, with prominent longitudinal ridges approximately 0·75 mm. apart and abundant closely adpressed 0·5–1 mm. long setae, often reddish tinged on ridges. Leaves sessile, coriaceous, sometimes turning red with age, narrowly obovate or elliptic, 15–60(–80) mm. long, 4–25 mm. wide, acute, sharply and rigidly mucronate, margins spinose, both surfaces bearing closely adpressed sparse to abundant setae, these often shorter, finer and less abundant on upper surface, arising from slight protuberances. Head 4–18-flowered, compact; bracts ovate-lanceolate or lanceolate, 8–17 mm. long, 3–6 mm. wide, acute, mucronate, margin spinose, surfaces bearing adpressed setae abundant on outer surface, sparse and finer on inner surface. Calyx with teeth up to 3 mm., prominently ribbed, 13–20 mm. long, 1–1·5 mm. wide, bearing closely adpressed setae between ribs. Corolla bright blue; tube 18–28 mm. long; lobes obcordate, apiculate, mucronate, 6–9 mm. long, 5–6 mm. broad. Seeds 6–7 mm. long, 2–2·5 mm. across. Fig. 2.

KENYA. Northern Frontier Province: Sololo, 7 Sept. 1952, *Gillett* 13789! & Burole, 17 Jan. 1972, *Bally & Smith* 14882!; Meru District: Isiolo, 15 Apr. 1944, *J. Bally* in *Bally* 3529!
DISTR. **K**1, 4; Sudan, Ethiopia, Somali Republic
HAB. Succulent evergreen bushland and semi-desert vegetation, often on rocky scarps; 700–1500 m.

SYN. *Valoradia abyssinica* Hochst. in Flora 25: 239 (1842)
 V. patula Hochst. in Flora 25: 240 (1842); Oliv., F.T.A. 3: 487 (1877). Type: Ethiopia, *Schimper* (? TUB, holo.)
 Ceratostigma speciosum Prain in J.B. 44: 8 (1906). Types: Somali Republic (N.), Golis, *Lort Philips* (BM, syn.!, K, isosyn.!) & Guldoo Hamud, *Cole* (BM, syn.!, K, isosyn.!) & Gan Libah, Habr Awal [Hadrawal], *Donaldson Smith* (BM, syn.!)

NOTE. An attractive plant that has been introduced into gardens locally, e.g. Nairobi, *Fosberg* 49860!

3. **PLUMBAGO**

L., Sp. Pl.: 151 (1753) & Gen. Pl., ed. 5: 75 (1754)

Perennial herbs or shrubs. Leaves alternate, simple. Flowers in alternate 3-bracteate 1-flowered spikelets grouped into elongated terminal spikes; central (outer) bract largest. Calyx tubular, 5-ribbed, scarious between ribs, 5-toothed. Corolla with narrow tube and rotate limb. Stamens free. Ovary 1-locular; style 1, with 5 stigma-lobes. Fruit a capsule, membranous, with 5 small apical reflexed lobes, 1-seeded, enclosed in the persistent calyx, dehiscent by a complete ring near base, often splitting into 5 valves from below. Seed dark brown or black, long, narrow, slightly flattened, tapering slightly towards apex, surface colliculate; hilum small, oval, in a longitudinal depression.

A genus of 24 species occurring almost throughout the world.

Two species, included in the following key, are widely cultivated.
 P. auriculata Lam. (*P. capensis* Thunb.; T.T.C.L.: 453 (1949); U.O.P.Z.: 417 (1949)) is commonly used as a hedge or ornamental plant. Perennial herb or small shrub to 2 m.; leaves shortly petiolate, obovate or elliptic; peduncle shortly and densely white-hairy; calyx 1–1·2 cm. long, with sparse short white hairs and fairly abundant stalked glands on the upper part; corolla light blue, the tube 25–30 mm. long, the lobes broadly obovate, 10–14 mm. long.

P. indica L. has been grown in Tanzania at Amani, e.g. *Greenway* 2236 ! Perennial to
60 cm.; leaves shortly petiolate, obovate or elliptic; peduncle glabrous; calyx 6–7 mm.
long, with sparse or abundant stalked glands; corolla bright scarlet, the tube 14–20 mm.
long, the lobes narrowly obovate, 9–11 mm. long.

1. Leaves, stems and bracts all hispid . . . 7. *P. ciliata*
   Leaves glabrous; stems glabrous, sessile-glandular
   or glandular-setose; bracts glabrous or sessile-
   glandular; no parts hispid 2
2. Corolla-lobes bright scarlet (cultivated) . . *P. indica*
                                                (see above)
   Corolla-lobes purple, magenta, blue or white . . . 3
3. Peduncles shortly and densely white-hairy (fig.
   3/8) 4
   Peduncles glabrous, sessile-glandular or glandular-
   setose, but without short white hairs 5
4. Corolla white; calyx with dense eglandular pubescence and short (0·25 mm. long), stout, abundant,
   evenly-distributed glandular setae (fig. 3/8b);
   leaves short-lived, up to 12 mm. long . . 8. *P. aphylla*
   Corolla light blue; calyx with sparse eglandular
   pubescence and longer (1 mm. long) more slender
   glandular setae, mostly on upper half; leaves
   persistent, more than 19 mm. long (cultivated) . *P. auriculata*
                                                (see above)
5. Peduncles glandular-setose (fig. 3/6a) 6
   Peduncles glabrous or with sessile glands 8
6. Corolla white; setae of calyx 1·5–2 mm. long (fig.
   3/6b); stem glandular-setose throughout its
   length . . . . 6. *P. glandulicaulis*
   Corolla blue, purple or magenta; setae of calyx up
   to 1 mm. long; stem never glandular below
   uppermost leaf 7
7. Corolla bright blue, large, the tube 1·7–2 times as
   long as the calyx and the lobes ± 8–10 × 6 mm. 4. *P. amplexicaulis*
   Corolla purple or magenta, small, the tube 1–1·3
   times as long as the calyx and the lobes ± 6 ×
   3 mm. . . . . 5. *P. montis-elgonis*
8. Bracts lanceolate, acuminate, very narrow, 5–10
   times as long as broad (fig. 3/3a) . . . 3. *P. stenophylla*
   Bracts various but relatively broader, 1–3 times as
   long as broad (fig. 3/1a, 2a) 9
9. Ridges of stem prominent, rather distant (1–2 per
   mm.); calyx-setae (fig. 3/1b) stout, rigid, straight;
   calyx-teeth up to 1·5 mm. long; corolla-lobes
   6–9 × 3–5 mm. . . . . 1. *P. zeylanica*
   Ridges of stem fine, closely-spaced (4 per mm.);
   calyx-setae (fig. 3/2b) fine, flexible, often curved;
   calyx-teeth 1·5–2 mm. long; corolla-lobes 9–13 ×
   5–6 mm. . . . . 2. *P. dawei*

1. **P. zeylanica** *L.*, Sp. Pl.: 151 (1753); Oliv., F.T.A. 3: 486 (1877);
Engl., P.O.A. C: 304 (1895); Wright in Fl. Cap. 4 (1): 425 (1906); Fries,
Wiss. Ergebn. Schwed. Rhod.-Kongo-Exped. 1: 254 (1916); T.S.K., ed. 2:
154 (1936); F.P.N.A. 2: 42 (1947); W.F.K.: 84 (1948); T.T.C.L.: 453
(1949); U.O.P.Z.: 417 (1949); van Steenis, Fl. Males., ser. 1, 4: 109 (1949);

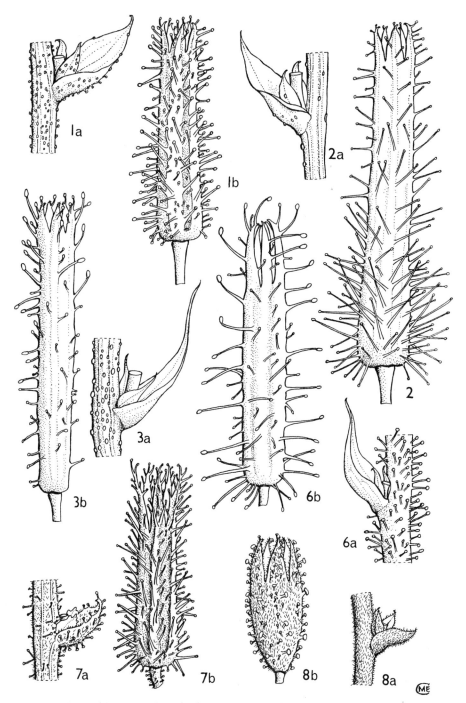

FIG. 3. *PLUMBAGO SPP.*—details of peduncle, bracts and pedicel (a), × 9, and calyx (b), × 6. Species numbered as in text. **1,** P. zeylanica, from *Gillett* 13570; **2,** P. dawei from *Fyffe* 87; **3,** P. stenophylla, from *R. M. Graham* A.615; **6,** P. glandulicaulis, from *Carmichael* 1340; **7,** P. ciliata, from *Busse* 2839; **8,** P. aphylla, from *Harris* 134.

F.P.S. 3: 68, fig. 12 (1956); Watt & Breyer-Brandwijk, Medic. Pois. Pl. S. & E. Afr., ed. 2: 850 (1962); F.F.N.R.: 318 (1962); Hepper, F.W.T.A., ed. 2, 2: 306, fig. 270 (1963); Dyer in Fl. S. Afr. 26: 17 (1963); Pohn., Roessler & Schreiber in Merxmüller, Prodr. Fl. S.W. Afr. 105: 4 (1967); Troupin, Syll. Fl. Rwanda: 202, fig. (1971); F.P.U., ed. 2: 116 (1971); U.K.W.F.: 580 (1974). Type: *Plumbago foliis petiolatis* L., Hort. Cliff.: 53 (1737) (lecto.!, BM, typolecto.!)

Creeping herb, scandent or semi-scandent shrub, laxly branched, 0·3–2·2 m. high. Stems wiry, tough, usually woody, glabrous, often with white waxy dots, bearing prominent longitudinal ridges spaced 1–2 per mm. Leaf-blades ovate, ovate-lanceolate, elliptic or oblong, rarely obovate, broad or narrow, 2·5–13 cm. long, 1–6 cm. wide, acute, acuminate or mucronate, base always cuneate, glabrous, often with white waxy dots on under surfaces; petiole 2–12 mm. long, base amplexicaul, sometimes auriculate. Peduncles bearing prominent often dense sessile glands. Flowers scented; pedicels 1–2 mm. long; bracts ovate, ovate-lanceolate or lanceolate, 2–8 mm. long, 1·5–2 mm. wide, prominently ribbed, bearing abundant evenly distributed stalked glands; stalks stout, rigid, straight, 1–2 mm. long. Calyx-teeth up to 1·5 mm. long. Corolla white; tube 17–26 mm.; lobes obovate, 6–9(–11) mm. long, 3–5 mm. broad, acute, with shortly excurrent central nerve. Seeds dark brown, 6 mm. long, 2 mm. wide. Fig. 3/1, p. 7.

UGANDA. W. Nile District: Nebbi, Sept. 1940, *Purseglove* 1062!; Teso District: Serere, June 1932, *Chandler* 676!; Mengo District: Kisinsi Point opposite Kaazi, 22 Feb. 1970, *Lye* 5070!
KENYA. Northern Frontier Province: Moyale, 10 July 1952, *Gillett* 13570!; Naivasha District: Hells Gates Gorge, Jan. 1960, *Seldon* in *E.A.H.* 11846!; Machakos District: Athi R. on main Nairobi–Mombasa road, 30 May 1958, *Verdcourt & Napper* 2163!; Teita District: Wusi–Mwatate road, 18 Sept. 1953, *Drummond & Hemsley* 4406!
TANZANIA. Maswa District: Moru Kopjes, 11 Apr. 1961, *Greenway & Turner* 10036!; Masai District: about 1·5 km. N. of Kwa Kuchinja, 21 July 1956, *Milne-Redhead & Taylor* 11281!; Iringa District: Mdonya R. crossing, 1 Apr. 1970, *Greenway & Kanuri* 14242!; Zanzibar I., Muyuni, 16 July 1933, *Vaughan* 2146!
DISTR. U1–4; K1–7; T1–8; Z; throughout the tropics and subtropics
HAB. Deciduous woodland, grassland and scrub, often by rivers and on lake margins; 0–2000 m.

2. **P. dawei** *Rolfe* in J.L.S. 37: 522 (1906); T.T.C.L.: 453 (1949); E.P.A.: 660 (1960). Type: Uganda, Toro District, Mahoma [Nsongi] R., *Dawe* 543 (K, holo.!)

Creeping or scandent herb or shrub, laxly branched, 0·9–5·7 m. Stems wiry, tough, usually woody, glabrous, bearing fine closely spaced longitudinal ridges, spaced 4 per mm. Leaf-blades ovate, ovate-lanceolate, elliptic or oblong, rarely orbicular, broad or narrow, 5·5–18 cm. long, 2·5–7 cm. wide, acute, acuminate or mucronate, base rounded or rarely cuneate, glabrous; petiole 1–2 cm. long, base amplexicaul, sometimes auriculate. Peduncle bearing sparse inconspicuous or sometimes prominent sessile glands, or glabrous. Flowers scented; pedicels 1–2 mm. long; bracts ovate, ovate-lanceolate or lanceolate, 2–8 mm. long, 1–2·5 mm. wide, acute, acuminate or mucronate, glabrous. Calyx (10–)15–20 mm. long, 1·5–2·5 mm. wide, seldom prominently ribbed, bearing abundant stalked glands; stalks thin, not rigid, often curved, 1·5–2 mm. long; calyx-teeth 1·5–2 mm. long. Corolla white; tube 24–27 mm.; lobes obovate, (9–)10–13 mm. long, 5–6 mm. broad, acute, with shortly excurrent central nerve. Fig. 3/2, p. 7.

UGANDA. Ankole District: Ruizi R., 15 Dec. 1950, *Jarrett* 39!; Busoga District: Chicot (? Chiko), Mar. 1916, *Dummer* 2782!; Mengo District: near Kigo, Mutungo, 17 Nov. 1969, *Lye* 4693!

KENYA. Northern Frontier Province: Mathews Range, 10 June 1959, *Kerfoot* 1128! & Marsabit, 18 Aug. 1959, *Ossent* 330!; Teita District: Voi area and in open patches of Lumi River Forest, Taveta, *Gardner* in *F.D.* 2968!

TANZANIA. Moshi District: Nanga R., 11 km. on Moshi–Himo road, 10 Aug. 1968, *Bigger* 2106!; Pare District: Gonjamaore, July 1955, *Semsei* 2120!; Morogoro District: Mtibwa Forest Reserve, Aug. 1952, *Semsei* 893!

DISTR. U2–4; K1, 7; T2, 3, 5–7; Ethiopia, Madagascar

HAB. Forest margins, clearings and secondary bushland, riverine forest; 600–2700 m.

SYN. *P. zeylanica* L. var. *dawei* (Rolfe) Mildbr., Z.A.E.: 518 (1913)

3. **P. stenophylla** *Wilmot-Dear* in K.B. 31 (1976). Type: Kenya, Kilifi District, Mida, *R. M. Graham* A. 615 in *F.D.* 2103 (EA, holo. !, BM, K, iso. !)

Woody herb or small shrub. Stems glabrous, bearing fairly prominent longitudinal ridges, spaced 1–4 per mm. Leaf-blades ovate-lanceolate, 4–8 cm. long, 1·5–3 cm. wide, acute, base cuneate, glabrous, rather membranous and almost translucent; petiole 5–10 mm. long, base amplexicaul, often auriculate. Peduncle bearing sparse or abundant, always prominent sessile glands; pedicels 1–2 mm. long; bracts lanceolate with a long acumen, very narrow, outer bract 4–5 mm. long, up to 1 mm. wide, inner pair 0·5–2 mm. long, up to 0·5 mm. wide, glabrous. Calyx 13–15 mm. long, 1–1·5 mm. wide, not prominently ribbed, bearing sparse or abundant evenly distributed stalked glands; stalks very thin, not rigid, often curved, 0·5–2 mm. long; calyx-teeth up to 1·5–2 mm. long. Corolla white; tube 26–28 mm.; lobes obovate, 10–11 mm. long, 5 mm. broad, acute, with shortly excurrent central nerve. Fig. 3/3, p. 7.

KENYA. Kilifi District: Mida, Sept. 1929, *R. M. Graham* A. 615 in *F.D.* 2103!; Lamu District: Boni Forest, Basuba, 7 Aug. 1975, *Katz* 75/48!

DISTR. K7; not known elsewhere

HAB. Coastal forest

4. **P. amplexicaulis** *Oliv.* in J.L.S. 15: 96 (1876) & F.T.A. 3: 487 (1877); Engl., P.O.A. C: 304 (1895); T.T.C.L.: 452 (1949); F.F.N.R.: 318 (1962). Type: Tanzania, Kigoma District, S. of Kawele, *Cameron* (K, holo. !)

Erect herb or subshrub, laxly branched, 45–90 cm. high; roots with terminal tuberous swellings; sap of stems and roots, but not tubers, yellow. Stems glabrous, often reddish, bearing prominent longitudinal ridges, 1–2 per mm. Leaves sessile or with short (10–20 mm.) winged petiole, amplexicaul, auriculate, auricles often large, glabrous, chartaceous; lower leaf-blades obovate or oblong, 5–24 cm. long, 2–11 cm. wide, upper often narrowly oblong or lanceolate, smaller, 4–11 cm. long, 1–2·5 cm. wide. Peduncles bearing abundant short hairs and dense stalked glands, often reddish; gland stalks up to 0·5 mm. long; pedicels 1 mm. long; bracts ovate-lanceolate or lanceolate, outer bract 2·5–8 mm. long, 1–2 mm. wide, acute, inner pair minute, less than 1 mm. long, usually glandular on under surface, both surfaces pilose, margins ciliate. Calyx 8–10 mm. long, 1·5–2 mm. wide, seldom prominently ribbed, bearing sparse or abundant stalked glands more numerous towards the apex; stalks very thin, not rigid, often curved, 0·5–1 mm. long; calyx-teeth up to 1·5 mm. long. Corolla-tube mauve or reddish, 15–27 mm. long, 1·7–2 times length of calyx; lobes bright deep blue, obovate to spathulate, 8–10 mm. long, 6 mm. broad, acute, with shortly excurrent central nerve. Fig. 4, p. 10.

TANZANIA. Ufipa District: Kasunga village, 14 Dec. 1958, *Richards* 10331!; Iringa District: Iheme, 23 Feb. 1962, *Polhill & Paulo* 1577! & Sao Hill road, 1 Feb. 1950, *Milton* 68!

DISTR. T4, 7; Zambia

HAB. Deciduous woodland and grassland, often on sandy or red soil; 900–2000 m.

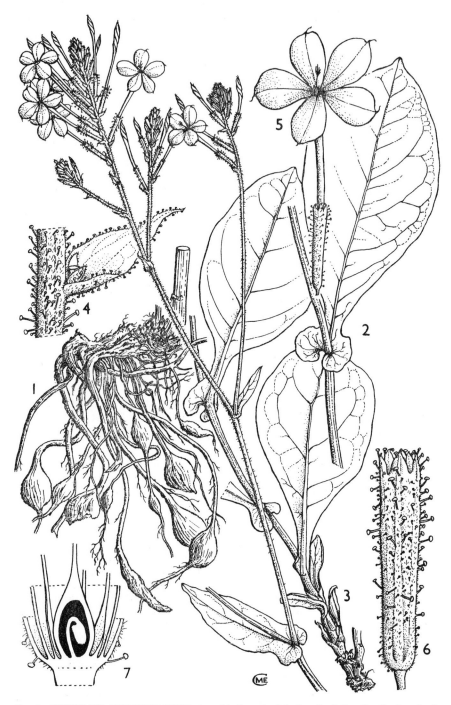

FIG. 4. *PLUMBAGO AMPLEXICAULIS*—**1**, rootstock, × ⅔; **2**, leaf, × ⅔; **3**, flowering shoot, × ⅔; **4**, detail of inflorescence-axis and bracts, × 8; **5**, flower, × 2; **6**, calyx, × 6; **7**, longitudinal section of ovary, × 20. 1, from *Polhill & Paulo* 1577; 2, from *Richards* 7071; 3–7, from *Richards* 11585.

5. **P. montis-elgonis** *Bullock* in K.B. 1932: 501 (1932); U.K.W.F.: 580 (1974). Type: Kenya, Mt. Elgon, *C. Lugard* 657 (K, holo.!, EA, iso.!)

Erect herb, laxly branched, up to 60 cm. high; sap profuse and colourless. Roots with terminal tuberous swellings. Stem glabrous, bearing prominent longitudinal ridges of varying width. Leaves sessile, amplexicaul, auriculate, the auricles often large, glabrous, chartaceous; lower leaf-blades obovate, 12–22 cm. long, 1–11 cm. wide, upper blades narrowly oblong, up to 9·5 cm. long, 2·5 cm. wide. Peduncles bearing small dense shortly-stalked glands, stalks up to 0·25 mm. long; pedicels up to 1 mm. long; bracts ovate-lanceolate, outer bract 1–3 mm. long, 0·5–2 mm. wide, inner pair minute, less than 1 mm. long, acute, surface glabrous, margins ciliate. Calyx often prominently ribbed, 11–13 mm. long, 1–1·5 mm. wide, bearing sparse evenly distributed stalked glands; stalks thin, not rigid, often curved, (0·25–)0·5–1·0 mm. long. Corolla magenta or purple; tube 13–14 mm. long, 1–1·3 times length of calyx; lobes obovate, up to 6 mm. long, 3 mm. broad, acute, with shortly excurrent central nerve.

KENYA. Trans-Nzoia District: Mt. Elgon, SE. slopes, 18 May 1931, *C. Lugard* 657!
TANZANIA. Ngara District: Bugufi, Ruganzo, 12 Dec. 1960, *Tanner* 5465!
DISTR. **K3**; **T1**; Ethiopia
HAB. Forest, sometimes riverine; 1350–1950 m.

6. **P. glandulicaulis** *Wilmot-Dear* in K.B. 31 (1976). Type: Tanzania, Mbulu District, Marang Forest, *Carmichael* 1340 (EA, holo.!)

Herb to 60 cm. high, laxly branched. Stems bearing prominent longitudinal ridges 1–2 mm. apart, and sparse stalked glands; stalks up to 1 mm. long. Leaves sessile, amplexicaul, auriculate; blades obovate (at least the upper leaves), 11–21 cm. long, 2·5–7 cm. wide, obtuse or acuminate, glabrous, somewhat membranous. Peduncles bearing dense stalked glands; stalks (0·5–)1 mm. long; pedicels up to 1 mm. long; bracts ovate-lanceolate, outer bract 3–4 mm. long, 1·5–2 mm. wide, inner pair minute, rarely up to 2 mm. long, 1 mm. wide, acute or acuminate, glabrous. Calyx 12 mm. long, 1·5 mm. wide, not prominently ribbed, dark coloured between ribs in basal half, bearing abundant evenly distributed stalked glands; stalks thin, not rigid, 1·5–2 mm. long, those arising from dark region shorter, stouter and darker; calyx-teeth up to 3 mm. long. Corolla white, tube approximately equalling calyx. Fig. 3/6, p. 7.

TANZANIA. Mbulu District: Marang Forest, 7 Jan. 1967, *Carmichael* 1340!
DISTR. **T2**; not known elsewhere
HAB. Forest; 1860 m.

7. **P. ciliata** *Wilmot-Dear* in K.B. 31 (1976). Type: Tanzania, Lindi District, *Busse* 2839 (EA, holo.!)

Herb to 45 cm. high. Stems bearing closely spaced longitudinal ridges, 2–4 per mm., sparse short weak hairs and abundant rigid hairs up to 2 mm. long. Leaves sessile, amplexicaul, obovate, up to 23 cm. long, 9 cm. wide, obtuse, somewhat chartaceous, with rigid hairs up to 2 mm. long arising from margins and from major and minor veins on both surfaces; midrib, major and minor veins very prominent on both surfaces. Peduncles bearing sparse short weak hairs and abundant small shortly stalked glands; stalks thin, not rigid, 0·25–1 mm. long; pedicels 1–2 mm. long, bearing stalked glands as on peduncles; outer bract ovate or ovate-lanceolate, 1–2 mm. long, 1–1·5 mm. wide, acute, bearing sparse short hairs on both surfaces and stalked glands beneath, inner pair broadly ovate with dentate margins, up to 1 mm. long and wide, scarious. Calyx 8–9 mm. long, 1·5–2 mm. wide, prominently ribbed, very densely covered with stalked glands; stalks stout, 1–1·25 mm. long, glands very small; calyx-teeth up to 3–4 mm. long. Corolla-tube 13–15 mm. long; lobes up to 9 mm. long, 8 mm. broad. Fig. 3/7, p. 7.

TANZANIA.   Lindi District: probably E. slopes Rondo [Muera] Plateau, 13 June 1903, *Busse* 2839!
DISTR.   **T8**; not known elsewhere
HAB.   *Brachystegia* woodland; ± 500 m.

8. **P. aphylla** *Boiss.* in DC., Prodr. 12: 694 (1848); van Steenis in Fl. Males., ser. 1, 4: 109 (1949). Type: Madagascar, *Bojer* (G, holo.)

Small shrub, virgate, scrambling or forming dense tangled clumps, up to 1 m. high. Stems wiry, pale yellow-green, glabrous, bearing numerous fine longitudinal ridges, 4 per mm. Leaves often soon disappearing, only the petiole bases remaining on stem as brown scales; leaf-blades obovate, 5–12 mm. long, 5–6 mm. broad, acute, glabrous, chartaceous; petiole winged, base amplexicaul. Peduncle bearing short dense white hairs; pedicels up to 0·5 mm. long. Bracts ovate or ovate-lanceolate, 1·5 mm. long, 1·5 mm. broad, acute, bearing short dense white hairs beneath. Calyx not prominently ribbed, 6–7 mm. long, 3 mm. wide, bearing short dense white hairs and abundant evenly distributed stalked glands; stalks stout, less than 0·25 (–0·5) mm., glands small; calyx-teeth up to 2 mm. long. Corolla white; tube 10–13 mm. long, 1·5–2 times length of calyx; lobes obovate, 6 mm. long, with shortly excurrent central nerve. Fig. 3/8, p. 7.

TANZANIA.   Uzaramo District: Mbudya I., 18 July 1965, *Harris* 134!
DISTR.   **T6**; Madagascar, Aldabra
HAB.   Near shore, on coral rock

SYN.   *P. parvifolia* Hemsley in J.B. 54: 362 (1916). Type: Aldabra, *Fryer* 115 (K, holo.!)

INDEX TO PLUMBAGINACEAE